Design and illustration by Hal Just

Library of Congress Cataloging in Publication Data

Simon, Seymour.
   Life and death in nature.

SUMMARY: Examines the functional purpose of death in the plant and animal world and shows how plants and animals serve as food for other animals and how plants absorb decayed matter.
   1. Death (Biology)—Juvenile literature. 2. Biogeochemical cycles—Juvenile literature. |1. Death. 2. Biogeochemical cycles| I. Just, Hal. II. Title.
QH530.S55   574.5   75-41496
ISBN 0-07-057456-1 lib. bdg.

Copyright © 1976 by Seymour Simon and Hal Just. All Rights Reserved. Printed in the United States of America. No part of this publication may be reproduced, stored in a retrieval system, or transmitted, in any form or by any means, electronic, mechanical, photocopying, recording, or otherwise, without the prior written permission of the publisher.

123456   RABP   789876

LIFE AND DEATH NATURE

BY SEYMOUR SIMON
ILLUSTRATED BY HAL JUST

Hoosic Valley Elementary Library
Schaghticoke, New York

McGraw-Hill Book Company
New York • St. Louis • San Francisco • Auckland • Düsseldorf
Johannesburg • Kuala Lumpur • London • Mexico • Montreal • New Delhi • Panama
Paris • São Paulo • Singapore • Sydney • Tokyo • Toronto

Life seems to awaken in the spring after a long winter rest. Elms, maples, and oaks are beginning to leaf and new shoots are pushing out of the ground. Winter-brown grasses on lawns are greening up.

After a spring shower,
you can smell the strong odor
of fresh earth.
Early-season daffodils and tulips
are showing their colors.

Animals are on the move, too.
Bees are searching for nectar in
newly opened flowers.
Ants and other insects
are awakening from the cold of winter
to the promise of warm weather in the air.

Robins and Jays are building their nests,
searching for food, mating and laying eggs.
In city parks and country fields
you can see squirrels chasing each other,
and maybe even a rabbit or two
munching along on a grassy patch of ground.

Spring is a good time
to explore the out-of-doors.
Go to a field, park, or vacant lot
where plants grow.
Pick up a handful of soil
and look at it carefully.
Do you see anything else
besides the new green shoots?

In the soil are old rotting leaves and stems,
dead insects, and the decaying parts of small animals.
What happens to the decaying plants and animals?
Rotting plants and animal parts change
into the <u>nutrients</u> that growing plants need.
Nutrients are minerals and other chemicals
that plants need for normal growth.

Plants grow better in soil where living things have lived and died.
When you look at the top of the soil you see only the plant parts that have just died and fallen to the ground.

Long-dead leaves, twigs, seeds, and animals
are buried deeper.
Dig down beneath the surface of the soil.
Look at a rotting leaf or a piece of wood.
Rub it between your fingers.
You can see that it has become soft and crumbly.

Insects, earthworms, and other soil animals eat the plant and animal parts in the ground.

Small non-green plants called <u>bacteria</u> and <u>fungi</u> often use dead animals and plants as food.

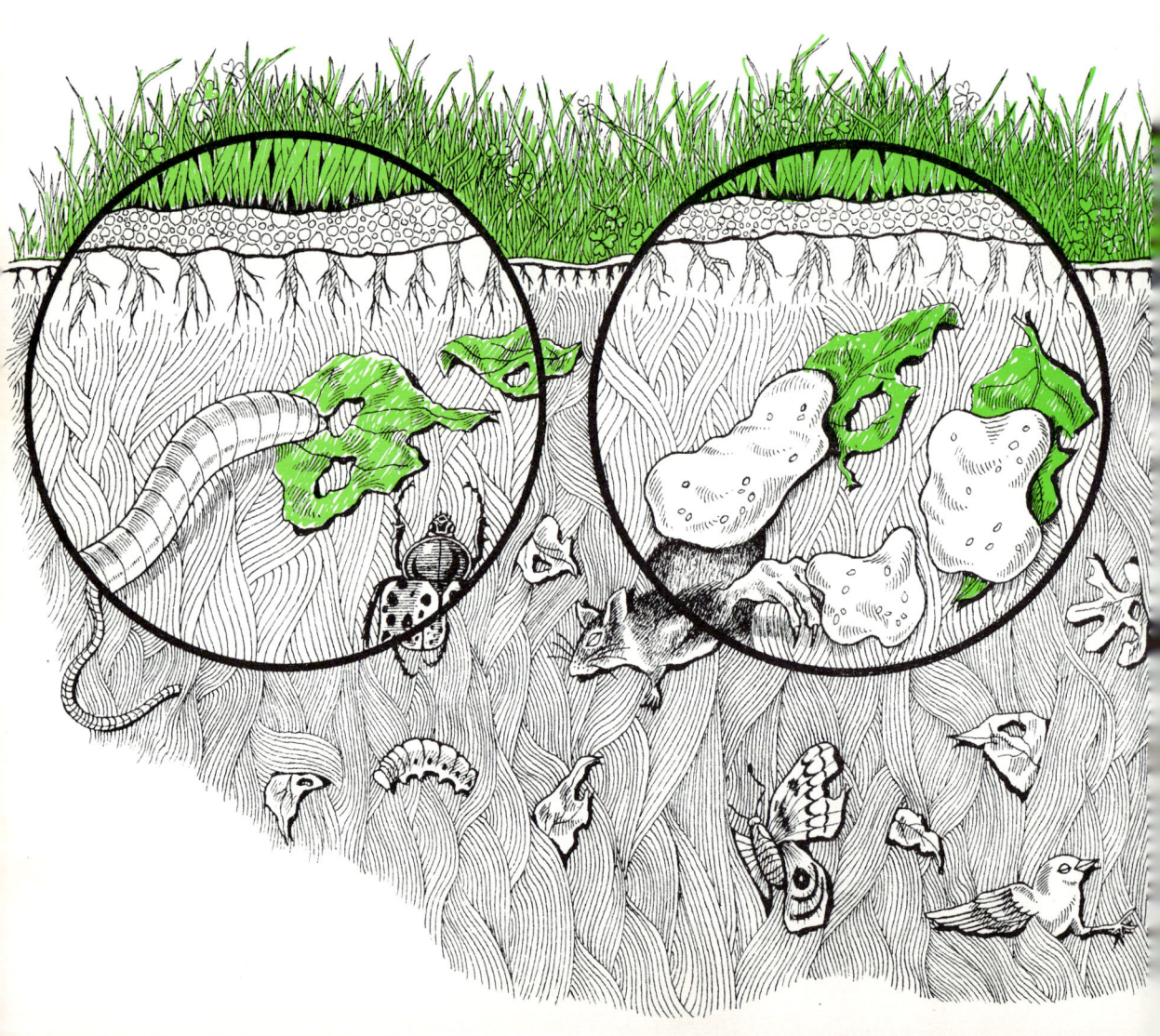

Nutrients from the
decaying plants and animals
are released into
the soil and the air.

Often
nutrients are used
by living things
once again.

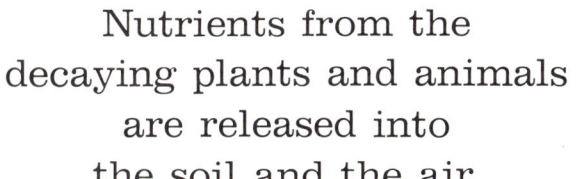

Only a green plant can make its own food.
Green plants have a very special substance within them.
The substance is called <u>chlorophyll</u>.
Light energy from the sun strikes the chlorophyll.
The chlorophyll helps the plants use the energy
to make food.

The food is made out of the minerals and other chemicals in the air, water, and soil around the plants.
So, green plants need the plant foods that dead plants and animals give back to their surroundings.

You can try an experiment at home to see how once-living things decay and change in the soil. You will need a plastic container or a coffee can, a hammer and a nail, soil, a leaf, a blade of grass, and water.

Use the hammer and the nail to punch a few holes around the side of the coffee can or the container.

The holes will let air circulate in the soil.

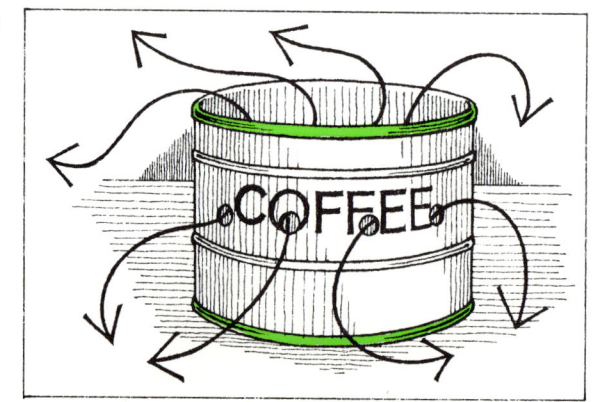

Place an inch or two of soil in the can.

Place the leaf and the blade of grass on top of the soil.

Add another one-inch layer of soil to cover them.

Add water so that the soil is damp but not soggy.

At the end of a week
move the top layer of soil to one side
and look at the leaf and the grass.
Have the leaf or the grass changed in color?
Do they have an odor?
Rub a bit of the leaf between your fingers.
Does it feel different from a living leaf?
Do you see any other changes taking place?

Replace the leaf and the grass and
again cover them with the soil.
Keep the soil moist as before.
Check the leaf and the grass again
for a second and a third week.
Keep a record of the changes you find.
How long does it take for the leaf and the grass
to show signs of decay?

Soft plant parts decay more quickly in the soil
than tougher stems or woody parts of a plant.
These may take many months to decay.
A fallen tree will take years to decay
and become part of the forest soil.
Smaller and softer plant and animal parts may decay
in just a few weeks.
Without death and decay in nature
there can be no new life.

New plants and animals
need the nutrients
that decay puts back
into the soil.

The journeys of nutrients
from living to nonliving things
are called <u>cycles</u>.
Nutrients keep moving in cycles
year after year after year.

A robin or an apple tree living today may contain nutrients that were once a part of a dinosaur, a rose petal, or an earthworm.

Nutrients are never used up, but passed on to each new living thing.

Can you imagine
some of the things that might happen
if living things never died?
For example, a pair of rabbits can have a litter
of half-a-dozen young every month or so.
In a few months, young rabbits are mature
and can have litters of their own.

If none died, there would be many millions of rabbits
at the end of two years.
At the end of ten years, the entire earth
would be completely buried in rabbits.
There would be no room for any more rabbits,
let alone any other living things.

Of course, something like that is not really possible.
Rabbits and other plant-eating animals
are themselves
a source of food for other animals.
Foxes, owls, snakes,
and many other kinds of animals
prey upon rabbits.
An increase in the number of rabbits
means an increase in the food supply
for rabbit eaters.

Too many rabbits in one place also means that there may not be enough plants for them to eat.
The lack of enough food makes some of the rabbits hungry, weak, and sick.
They fall easy prey to the rabbit eaters.
After a while, the number of rabbits begins to drop.
In nature, the numbers of one kind of living thing depends upon the numbers of other kinds of living things.
A balance of life is usually found.

You can experiment at home to see how the balance of life is usually kept in nature.

You will need three wide-mouthed glass jars, covers, soil, leaves and grasses, water, and a large number of earthworms.

You can usually collect earthworms under rotting logs or beneath piles of decaying leaves.

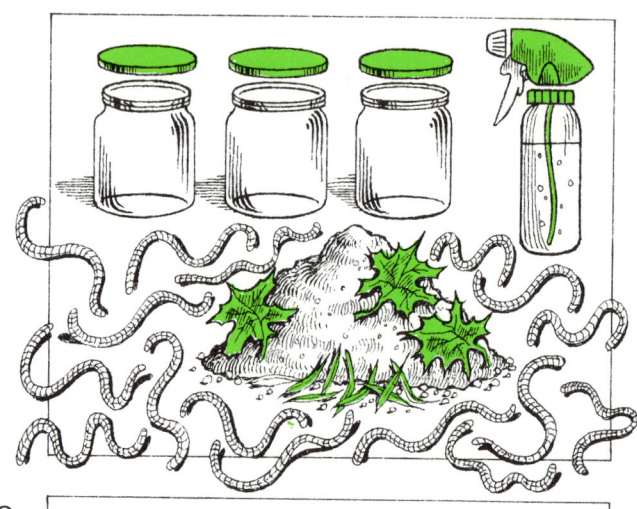

At home, wash and clean the jars carefully.

Place some soil and plant material in the bottom of each jar.

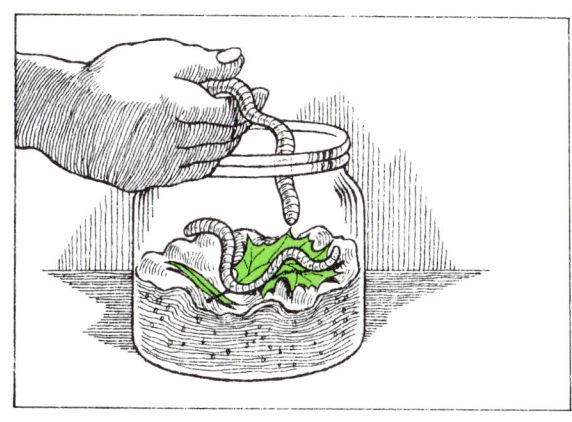

In the first jar, place one or two earthworms.

In the second jar, place five or six earthworms.

In the third jar, place ten or more earthworms.

Keep the soil in each jar moist but not soggy. Loosely cover the jars and keep them out of direct sunlight.

Check on the earthworms in the jars each day.
What do they seem to be doing?
Do the earthworms stay in one part of the soil in the jar,
or are they spread through the soil?
Is anything happening to the leaves and grasses
in the jars?

Place one of the earthworms from each of the jars
on a piece of wet paper towel.
Do they twist and wriggle all alike,
or are some more active than others?
Can you see any differences within the jars
after a few days?
After a week or two?
Do the number of earthworms in the jars stay the same?
How can you explain what is happening?

Have you ever kept fish in a home aquarium? What do you think happens if you try to keep twenty or thirty or a hundred fish in a small aquarium?

Many become sick and die.
After a few days, only a few fish
are left alive.

The same kind of thing happens
if you plant too many seeds in a small flower pot.
Many seeds will not sprout.
Those that do will be weak and spindly.
Many seedlings will die before
they are full grown.

All animals and plants need food and a certain amount of room to live and grow. In a balanced aquarium, you keep the numbers of animals and plants from growing too large for the tank. You make sure that each animal gets enough food. In a garden, you plant the seeds far enough apart so that each seedling will have room to grow.

Much the same thing happens in nature. Too many animals or plants cannot live in the same place. They crowd each other out. Some become sick and die. The ones that live use the dead animals and plants as food.

So, nothing is wasted in nature. Each plant and animal, dead or alive, becomes part of the endless chain of life.

about the author

Seymour Simon is the author of over forty science books for children. Many of his books have been selected by the National Science Teachers Association as Outstanding Science Trade Books for Children. These include DISCOVERING WHAT GARTER SNAKES DO and DISCOVERING WHAT CRICKETS DO from the "Discovering What Animals Do" series, and ABOUT YOUR HEART and HOT AND COLD from a series called "Let's Try It Out."

Mr. Simon has taught science at many levels in the New York City public school system. He and his wife and two boys live in Great Neck, New York.

about the artist

Hal Just nests in a studio apartment on the east side of New York. As a kid growing up in Brooklyn, life centered around playing stick ball, painting and drawing. He still paints and draws, and has been an art director for over twenty years.

Not long ago, while on a holiday, Mr. Just started a series of line drawings which developed into his first freelance, illustrative portfolio. Commissions followed from newspapers, national magazines and ad agencies. This is Hal Just's first book.